Energy Sector Standard of the People's Republic of China

NB/T 10349-2019
Replace DL/T 709-1999

Technical code for safety inspection of steel penstocks

压力钢管安全检测技术规程

(English Translation)

China Water & Power Press
中国水利水电出版社
Beijing 2024

All rights reserved. No part of this publication may be reproduced, stored in a retrieval system, or transmitted in any form or by any means—electronic, mechanical, photocopying, recording or otherwise, without prior written permission of the publisher.

图书在版编目（CIP）数据

压力钢管安全检测技术规程 : NB/T 10349-2019 = Technical code for safety inspection of steel penstocks (NB/T 10349-2019) : 英文 / 国家能源局发布. -- 北京 : 中国水利水电出版社, 2024. 7. ISBN 978-7-5226-2602-4

I. U173.1-65

中国国家版本馆CIP数据核字第2024HP0500号

Energy Sector Standard of the People's Republic of China

中华人民共和国能源行业标准

Technical code for safety inspection of steel penstocks

压力钢管安全检测技术规程

NB/T 10349-2019

Replace DL/T 709-1999

(English Translation)

Issued by National Energy Administration of the People's Republic of China
国家能源局　发布
Translation organized by China Renewable Energy Engineering Institute
水电水利规划设计总院　组织翻译
Published by China Water & Power Press
中国水利水电出版社　出版发行
　　Tel: (+ 86 10) 68545888　68545874
　　sales@mwr.gov.cn
　　Account name: China Water & Power Press
　　Address: No.1, Yuyuantan Nanlu, Haidian District, Beijing 100038, China
　　http://www.waterpub.com.cn
中国水利水电出版社微机排版中心　排版
北京中献拓方科技发展有限公司　印刷
210mm×297mm　16开本　1.75印张　70千字
2024年7月第1版　2024年7月第1次印刷

Price（定价）**:￥290.00**

About English Translation

This English version is one of China's energy sector standard series in English. Its translation was organized by China Renewable Energy Engineering Institute authorized by National Energy Administration of the People's Republic of China in compliance with relevant procedures and stipulations. This English version was issued by National Energy Administration of the People's Republic of China in Announcement [2023] No. 8 dated December 28, 2023.

This version was translated from the Chinese Standard NB/T 10349-2019, *Technical code for safety inspection of steel penstocks*, published by China Water & Power Press. The copyright is reserved by National Energy Administration of the People's Republic of China. In the event of any discrepancy in the implementation, the Chinese version shall prevail.

Many thanks go to the staff from relevant standard development organizations and those who have provided generous assistance in the translation and review process.

For further improvement of the English version, any comments and suggestions are welcome and should be addressed to:

China Renewable Energy Engineering Institute
No. 2 Beixiaojie, Liupukang, Xicheng District, Beijing 100120, China
Website: www.creei.cn

Translating organizations:

Hohai University

POWERCHINA Kunming Engineering Corporation Limited

Translating staff:

QIN Zhansheng	JIA Haibo	XIA Shifeng	BU Xiangang
YIN Guojiang	ZHENG Shengyi	WANG Shuo	ZHU Hai
LU Yulin			

Review panel members:

LIU Xiaofen	POWERCHINA Zhongnan Engineering Corporation Limited
LIANG Hongli	Shanghai Investigation, Design & Research Institute Co., Ltd.
LIU Qing	POWERCHINA Northwest Engineering Corporation Limited
QI Wen	POWERCHINA Beijing Engineering Corporation Limited
HU Baowen	POWERCHINA Huadong Engineering Corporation Limited
WANG Jingkun	POWERCHINA Huadong Engineering Corporation Limited
LIN Zhaohui	China Renewable Energy Engineering Institute
LI Shisheng	China Renewable Energy Engineering Institute

National Energy Administration of the People's Republic of China

翻译出版说明

本译本为国家能源局委托水电水利规划设计总院按照有关程序和规定，统一组织编译的能源行业标准英文版系列译本之一。2023年12月28日，国家能源局以2023年第8号公告予以公布。

本译本是根据中国水利水电出版社出版的《压力钢管安全检测技术规程》NB/T 10349—2019翻译的，著作权归国家能源局所有。在使用过程中，如出现异议，以中文版为准。

本译本在翻译和审核过程中，本标准编制单位及编制组有关成员给予了积极协助。

为不断提高本译本的质量，欢迎使用者提出意见和建议，并反馈给水电水利规划设计总院。

地址：北京市西城区六铺炕北小街2号
邮编：100120
网址：www.creei.cn

本译本编译单位：河海大学
中国电建集团昆明勘测设计研究院有限公司

本译本编译人员：秦战生　贾海波　夏仕锋　卜现港
尹国江　郑圣义　王　硕　朱　海
陆玉琳

本译本审核人员：

刘小芬　中国电建集团中南勘测设计研究院有限公司
梁洪丽　上海勘测设计研究院有限公司
柳　青　中国电建集团西北勘测设计研究院有限公司
齐　文　中国电建集团北京勘测设计研究院有限公司
胡葆文　中国电建集团华东勘测设计研究院有限公司
王靖坤　中国电建集团华东勘测设计研究院有限公司
林朝晖　水电水利规划设计总院
李仕胜　水电水利规划设计总院

国家能源局

Contents

Foreword		VII
1	Scope	1
2	Normative references	1
3	Basic requirements	1
3.1	Inspection body, personnel and equipment	1
3.2	Inspection content and items	2
3.3	Inspection interval	2
3.4	Technical information	2
4	Patrol inspection	3
5	Appearance inspection	4
6	Material testing	5
7	Corrosion inspection	5
8	Nondestructive testing	6
9	Stress detection	7
9.1	Detection requirements	7
9.2	Layout of detection points	7
9.3	Static stress detection	8
9.4	Dynamic stress detection	8
10	Vibration detection	8
10.1	Detection requirements	8
10.2	Detection instruments	8
10.3	Detection points layout	9
10.4	Detection of vibration response	9
10.5	Detection of dynamic characteristics	10
11	Check calculation	10
12	Safety evaluation	10
13	Inspection report	11
Annex A (informative) Patrol inspection record form		12
Annex B (informative) Appearance inspection record form		16

Foreword

This standard is drafted in accordance with the rules given in the GB/T 1.1-2009, *Directives for standardization—Part 1: Structure and drafting of standards.*

This standard replaces DL/T 709-1999, *Technical code for safety inspection of steel penstocks*. Besides editorial changes, the main technical revisions are as follows:

—Adding the content of corrosion degree assessment (see Clause 7).

—Supplementing the content of time of flight diffraction (TOFD) inspection, and revising the detection ratio of welds (see 8.4 and 8.6).

—Supplementing and improving the content of stress detection (see Clause 9, which is Clause 8 in DL/T 709-1999).

—Supplementing and improving the content of vibration detection (see Clause 10, which is Clause 9 in DL/T 709-1999).

—Deleting the content of water quality and sediment inspection (see Clause 10 in DL/T 709-1999).

—Adding the content of check calculation (see Clause 11).

National Energy Administration of the People's Republic of China is in charge of the administration of this standard. China Renewable Energy Engineering Institute has proposed this standard and is responsible for its routine management. Energy Sector Standardization Technical Committee on Hydropower Steel Structures and Hoists is responsible for the explanation of specific technical contents. Comments and suggestions in the implementation of this standard should be addressed to:

China Renewable Energy Engineering Institute
No. 2 Beixiaojie, Liupukang, Xicheng District, Beijing 100120, China

Drafting organization:

Hohai University

Chief drafting staff:

ZHENG Shengyi	XIA Shifeng	BU Xiangang	QIN Zhansheng
YE Huashun	LIU Weidong	ZHANG Chenghua	

The previous edition replaced by this standard:

—DL/T 709-1999

Technical code for safety inspection of steel penstocks

1 Scope

This standard specifies the content and technical requirements for the safety inspection of in-service steel penstocks in hydropower projects.

This standard is applicable to the safety inspection of in-service steel penstocks in hydropower projects.

2 Normative references

The following referenced documents are indispensable for the application of this document. For dated references, only the edition cited applies. For undated references, the latest edition of the referenced document (including all amendments) applies.

GB/T 1172, *Conversion of hardness and strength for ferrous metal*

GB/T 3323, *Radiographic examination of fusion welded joints in metallic materials*

GB/T 11345, *Non-destructive testing of welds—Ultrasonic testing—Techniques, testing levels, and assessment*

GB/T 29712, *Non-destructive testing of welds—Ultrasonic testing—Acceptance levels*

GB 50766, *Code for manufacture installation and acceptance of steel penstocks in hydroelectric and hydraulic engineering*

NB/T 35056, *Design code for steel penstocks of hydroelectric stations*

NB/T 47013.4, *Nondestructive testing of pressure equipments—Part 4: Magnetic particle testing*

NB/T 47013.5, *Nondestructive testing of pressure equipments—Part 5: Penetrant testing*

NB/T 47013.10, *Nondestructive testing of pressure equipments—Part 10: Ultrasonic time of flight diffraction technique*

3 Basic requirements

3.1 Inspection body, personnel and equipment

3.1.1 The inspection body shall have the qualification certificate on inspection and testing issued by the national supervisory and regulatory authority, and the inspection products or categories, inspection items or parameters, and inspection scope authorized by the certificate shall meet the requirements for safety inspection of steel penstocks.

3.1.2 Nondestructive testing (NDT) personnel shall have the qualification certificate corresponding to their job content issued by the Chinese Society for NDT or the hydropower sector. NDT results shall be evaluated by the NDT personnel with a qualification certificate of Level 2 or above.

3.1.3 The inspection personnel shall fully understand the design, manufacture, installation and operation of steel penstocks, have a good professional knowledge, and be familiar with inspection and testing methods.

3.1.4 The precision of the instruments and equipment used for inspection shall meet the requirements and shall be verified or calibrated by a metrological verification institution at or above municipal level.

3.2 Inspection content and items

3.2.1 The safety inspection of steel penstocks shall include:

a) Field inspection;

b) Check calculation;

c) Safety evaluation.

3.2.2 Field inspection shall include:

a) Patrol inspection;

b) Appearance inspection;

c) Material testing;

d) Corrosion inspection;

e) NDT;

f) Stress detection;

g) Vibration detection.

3.3 Inspection interval

3.3.1 The safety inspection of steel penstocks shall be conducted regularly. The inspection interval may be determined according to the service time and running conditions of steel penstocks. Safety inspections should include initial inspection, regular inspection and special inspection.

3.3.2 The initial inspection shall be carried out within 5 years after steel penstocks are put into operation, and the inspection items shall comply with 3.2.2.

3.3.3 After the initial inspection, a regular inspection shall be carried out for steel penstocks every 5 to 10 years, and the inspection items may be selected according to the running status of steel penstocks.

3.3.4 Special inspection shall be performed immediately in any of the following cases:

a) Steel penstocks are subjected to irresistible natural disasters, operation beyond design operating conditions, quality or safety accidents, etc. during operation;

b) Hazardous defects affecting the safety of steel penstocks are identified during operation;

c) The running condition of steel penstocks is obviously abnormal, which might affect the operation safety of the project.

3.3.5 For special inspection, patrol inspection and appearance inspection shall be carried out first, and other items may be inspected when necessary.

3.3.6 The steel penstocks shall be thoroughly inspected before upgrading of generating units.

3.4 Technical information

3.4.1 The operator of steel penstocks shall provide the inspection body with relevant technical information required for safety inspection.

3.4.2 The following information shall be collected before safety inspection:

a) As-built drawings and relevant information;

b) Quality certificate and re-inspection reports of main materials;

c) Inspection records, test records and relevant information on manufacture and installation;

d) Nondestructive testing reports of welds;

e) Corrosion inspection reports;

f) Major defect treatment records and relevant meeting minutes;

g) Acceptance reports of manufacture and installation quality;

h) Design amendment notifications and relevant design data;

i) Reports (records) of operation, maintenance and inspection;

j) Monitoring data and analysis reports;

k) Previous safety inspection and evaluation reports.

4 Patrol inspection

4.1 The inspection body shall carry out on-site patrol inspection according to the operation management information such as the patrol inspection records of steel penstocks provided by the operator. See Annex A for the patrol inspection record form.

4.2 The patrol inspection of exposed penstocks shall include:
   a) Deformation and displacement of pipe body;
   b) Seepage and stroke of expansion joints;
   c) Sealing performance of manholes, other openings, and pipe joints;
   d) Running status of vents or air valves;
   e) Running status of drainage facilities;
   f) Displacement and settlement of support piers and anchor blocks;
   g) Movability and lubrication of supports;
   h) Sealing performance of the connection between inlet valve and steel penstock.

4.3 The patrol inspection of embedded penstocks shall include:
   a) Leakage along the embedded penstocks;
   b) Running status of groundwater drainage facilities;
   c) Sealing performance of manholes, other openings, and pipe joints;
   d) Running and damage status of relevant monitoring instruments.

4.4 The patrol inspection of penstocks embedded in dam shall include:
   a) Cracking of downstream dam surface;
   b) Seepage of downstream dam surface;
   c) Displacement and deformation of downstream dam surface.

4.5 The patrol inspection of steel-lined reinforced concrete penstocks shall include:
   a) Circumferential cracks in the upper bend section;
   b) Circumferential cracks in the oblique straight section;
   c) Axial cracks in the oblique straight section;

d) Seepage and stroke of expansion joints;

e) Sealing performance of manholes, other openings and pipe joints;

f) Running and damage status of relevant monitoring instruments.

5 Appearance inspection

5.1 Before appearance inspection, the data on the manufacture, installation, and acceptance of steel penstocks, the treatment of major defects, the operation, repair, maintenance and various anomalies occurring during operation shall be acquainted.

5.2 The appearance inspection shall focus on the repaired parts during manufacture and installation, and the abnormal parts of steel penstocks during operation.

5.3 The results of appearance inspection shall be recorded in time. The appearance inspection record form is in Annex B. When necessary, auxiliary means, such as video and photograph, may be used to record and describe the inspection results.

5.4 For appearance inspection, measuring tools and apparatus such as level, ruler, caliper, tape measure, feeler gauge and weld inspection ruler should be used.

5.5 The appearance inspection shall include the pipe body, supporting rings and supports, expansion joints, and connections.

5.6 The appearance inspection of pipe body shall include:

a) Deformation of pipe wall such as indentation and bulging;

b) Corrosion of pipe wall;

c) Surface defects of main load-bearing welds;

d) Damage and deformation of stiffener rings;

e) State of grouting holes and their surrounding areas;

f) Damage and deformation of manhole covers.

5.7 The appearance inspection of supporting rings and supports shall include:

a) Deformation and damage of supporting rings and supports;

b) Running and contact status of supports;

c) Corrosion of supporting rings and supports.

5.8 The appearance inspection of expansion joints shall include:

a) Deformation and damage;

b) Running status;

c) Corrosion status.

5.9 The appearance inspection of connections shall include:

a) Appearance of connections between the penstock and concrete structures such as anchor block, dam and wall;

b) Connection between the penstock and the stiffener ring, supporting ring, or expansion joint;

c) Connection between the penstock and the inlet valve;

d) Cavitation of concrete around penstocks.

6 Material testing

6.1 Material testing may not be carried out when the material quality certificate and the documents of manufacture, installation and acceptance can prove that the material quality meets the design requirements.

6.2 Material testing shall be carried out to determine the material grade and properties when the material grade is not clear or doubtful.

6.3 When field conditions permit, the materials shall be sampled to analyze the mechanical properties and chemical composition so as to determine the material grade and properties. The sampling points shall be located in the less stressed parts which are easy to repair, and repair measures shall be proposed in advance. In sampling, acute angle shall be avoided. Instead, circular arc transition shall be adopted, and the arc radius shall not be less than 3 times the plate thickness and not be less than 30 mm.

6.4 When it is impossible to sample in the field, the chemical composition of the material may be analyzed by a spectrum analyzer or drilling out some chips on the part less stressed, and the material hardness shall be measured and converted into material strength in accordance with GB/T 1172, so as to determine the material grade and properties.

6.5 When quality problems affecting the operation safety of penstocks are found, samples shall be taken directly on penstocks for mechanical property test, chemical composition analysis, and metallographic analysis to determine the material grade and properties.

6.6 When the penstocks fail, samples shall be taken on the damaged pipe section for mechanical property test, chemical composition analysis, and metallographic analysis to determine the material grade and properties.

7 Corrosion inspection

7.1 Corrosion inspection shall include corrosion status inspection and corrosion measurement.

7.2 Corrosion status inspection should use measuring tools such as tape measure and ruler. For corrosion measurement, measuring instruments and tools such as thickness gauge, depthometer, and depth vernier caliper should be used.

7.3 Corrosion inspection should be carried out in sections, including upper horizontal section, inclined section, lower horizontal section and other sections, or in segments according to the circumferential welds of the penstock.

7.4 Before corrosion measurement, the surface of the inspected section shall be cleaned to remove the attachments, sludge, rust, etc.

7.5 Corrosion measurement shall meet the following requirements:
  a) In different penstock sections, those segments with a relatively high corrosion degree shall be tested, and no less than 30 % of the total segments shall be tested;
  b) On a selected pipe segment, at least 3 cross sections shall be measured. Each cross section shall be zoned according to the distribution of longitudinal welds, and the measuring points in each zone shall not be less than 3;
  c) For the penstock section or segment with a relatively high corrosion degree, the number of measuring segments and points should be increased appropriately;

d) The coating should be removed prior to inspection; otherwise, the coating thickness shall be deducted.

7.6 The corrosion degree shall be evaluated and classified into Grades A, B, C and D based on the results of corrosion status inspection and corrosion measurement. The evaluation criteria shall be as follows:

a) Grade A refers to slight corrosion. The coating shall be basically intact. There are only a few shallow and scattered corrosion pits on the surface of penstocks, and there are no more than 3 corrosion pits within an area of 300 mm × 300 mm;

b) Grade B refers to moderate corrosion. Local peeling of coating appears. There are no more than 30 corrosion pits within an area of 300 mm × 300 mm on the surface of penstocks. The average depth of corrosion pits is less than 5 % of the wall thickness and no more than 1.0 mm. The maximum depth of corrosion pits is less than 10 % of the wall thickness and no more than 2.0 mm;

c) Grade C refers to severe corrosion. The coating peels off in large areas, and the peeling area is greater than 100 mm × 100 mm; there are intensive corrosion pits on the surface of penstocks, the area of corrosion pits shall be less than 10 % of the surface area of the segment; the number of corrosion pits is more than 30 in an area of 300 mm × 300 mm; the average depth of corrosion pits is less than 10 % of the wall thickness and no more than 2.0 mm; the maximum depth of corrosion pits is less than 15 % of the wall thickness and no more than 3.0 mm;

d) Grade D refers to very severe corrosion. There are relatively deep and intensive corrosion pits, some of which are very deep; the area of intensive corrosion pits shall be less than 10 % of the surface area of the segment. The average depth of corrosion pits is more than 10 % of the wall thickness or more than 2.0 mm. The maximum depth of corrosion pits is more than 15 % of the wall thickness or more than 3.0 mm.

7.7 The corrosion inspection results shall include:

a) The characteristics, positions, distribution and area of the corrosion of each penstock section or segment, and the percentage of corrosion area in the area of penstock section or segment;

b) The corrosion and frequency distribution of each penstock section or segment, and the average corrosion, maximum corrosion, average corrosion rate and maximum corrosion rate of each penstock section or segment;

c) The average corrosion, maximum corrosion, average corrosion rate and maximum corrosion rate of the local area where the corrosion degree of each penstock section or segment is Grade D.

8 Nondestructive testing

8.1 Class Ⅰ and Class Ⅱ welds shall be subjected to NDT. The classification of welds shall comply with GB 50766.

8.2 Before NDT, the attachments, dirt, corrodes, etc. on and near the surface of welds shall be removed, and the surfaces on both sides of the welds should be ground if necessary.

8.3 Magnetic particle testing (MT) or penetrant testing (PT) may be used for the surface quality inspection of welds. MT shall comply with NB/T 47013.4; PT shall comply with NB/T 47013.5. For both MT and PT, the acceptance shall be Level Ⅰ.

8.4 Ultrasonic testing (UT) or radiographic testing (RT) may be used for the internal quality inspection of welds, and shall meet the following requirements:

   a) UT may adopt the pulse reflection method and TOFD technique. UT by pulse reflection method shall comply with GB/T 11345, and its testing level shall be Level B; its evaluation shall comply with GB/T 29712, and its acceptance level shall be Level 2. UT by TOFD shall comply with NB/T 47013.10, and its acceptance level shall be Level Ⅱ;

   b) RT shall comply with GB/T 3323. For Class Ⅰ welds, Level Ⅱ is deemed qualified. For Class Ⅱ welds, Level Ⅲ is deemed qualified.

8.5 When the defect found cannot be identified either qualitatively or quantitatively by a certain NDT method, other NDT methods shall be used for retesting. When different NDT methods are used for testing the same welding position or welding defect, the weld quality level shall be evaluated respectively and meet the corresponding requirements.

8.6 For the welds of different classes, the percentage of testing length to the total length shall meet the following requirements:

   a) Class Ⅰ welds: the testing length percentage shall not be less than 10 % for UT, not be less than 5 % for TOFD, and not be less than 2 % for RT;

   b) Class Ⅱ welds: the testing length percentage shall not be less than 5 % for UT, not be less than 3 % for TOFD, and not be less than 1 % for RT;

   c) When a defect exceeding limits like a crack is found in a weld, the entire weld shall be tested;

   d) In the case that multiple defects are found in the weld, the testing percentage shall be increased.

8.7 The number of welds for testing shall meet the following requirements:

   a) For Class Ⅰ welds, the number shall not be less than 20 % of the total;

   b) For Class Ⅱ welds, the number shall not be less than 10 % of the total.

8.8 The position with defects exceeding limits found previously or repaired shall be tested by 100 % with the previous testing method.

8.9 For cracks or other defects exceeding limits found in NDT, the cause shall be analyzed, the development trend shall be predicted, the severity of defects shall be evaluated, and treatment suggestions shall be proposed.

9 Stress detection

9.1 Detection requirements

9.1.1 Stress detection should be conducted for exposed penstock sections, but not for embedded sections.

9.1.2 The strain gaging method should be used for stress detection.

9.1.3 Information such as upstream and downstream water levels, working conditions and running time shall be recorded during stress detection.

9.2 Layout of detection points

9.2.1 Prior to stress detection, the stress of penstock structure should be calculated and analyzed in accordance with NB/T 35056 to determine the distribution and number of detection points.

9.2.2 The layout of detection points shall meet the following requirements:

   a) The detection points shall be representative, and shall be mainly arranged in highly stressed area, complex stress area and severely corroded area;

   b) For a symmetrical structure, the detection points may be arranged on one side, but an appropriate number of detection points shall be arranged on the other side for comparison;

   c) To check the reliability of detection data, an appropriate number of detection points for checking shall be arranged.

9.2.3 Detection points shall be arranged at these positions, such as the pipe wall in the midspan of penstocks, the pipe wall near the supporting ring, the pipe wall near the stiffener ring, the pipe wall near the connection of steel pipe and expansion joint, the pipe wall near the connection of the steel pipe and concrete structures like the anchor block, support pier, dam, and wall.

9.2.4 The sensing elements shall be firmly fixed to the pipe wall, insulated and moisture-proof. When a sensing element works under water, it shall be subjected to waterproof treatment. The signal transmission line shall be properly fixed and the resistance value shall be stable. When the resistance of the signal transmission line affects the measurement results, the measurement results shall be corrected.

9.3 Static stress detection

9.3.1 Static stress detection should be carried out under the design or quasi design condition.

9.3.2 The static stress detection should be repeated for 2 to 3 times.

9.3.3 If the difference of static stress values is more than 10 %, the cause shall be analyzed and the detection shall be repeated.

9.3.4 The detection results shall be compared with the calculation results of the corresponding working conditions. The stress values of design condition and check condition may be estimated according to the stress value of corresponding working condition if necessary.

9.4 Dynamic stress detection

9.4.1 Dynamic stress detection should be combined with load rejection test of the generating unit.

9.4.2 The working conditions for dynamic stress detection should include 25 %, 50 %, 75 % and 100 % rated load rejection.

9.4.3 The data of dynamic stress detection shall be continuously collected to obtain the complete stress-strain curve.

10 Vibration detection

10.1 Detection requirements

10.1.1 Vibration detection shall be carried out when vibration occurs during operation of the penstock and might affect the operation safety.

10.1.2 Vibration detection shall cover vibration response parameters, such as displacement, velocity and acceleration; and dynamic characteristics, such as natural frequency, damping ratio, and vibration mode.

10.2 Detection instruments

10.2.1 Detection instruments for penstocks shall adapt to the on-site conditions, including temperature, humidity, noise, etc.

10.2.2 For detection instruments, the measuring frequency range shall cover the useful frequency range of the measured signal, and the measuring dynamic range shall adapt to the variation range of the signal. The sampling frequency of the detection instrument shall not be less than 5.12 times the highest signal frequency, and anti-aliasing filtering shall be considered before sampling. The overall resolution of the detection instrument shall reach 10 μm for measuring vibration displacement.

10.2.3 The measuring frequency range of sensor shall meet the requirements of vibration frequency detection. Before vibration detection, the vibration frequency range of the tested object may be estimated to select the corresponding sensors.

10.2.4 Displacement, velocity and acceleration should be measured by dedicated sensors. When dedicated sensors are unavailable, the required vibration parameters may be obtained by integrating or differentiating the sensor output, and the possible errors shall be eliminated.

10.2.5 The sensitivity of the sensor shall ensure that the minimum measured signal level is 10 dB greater than the dynamic range lower limit level of the measuring system. The sensitivity of the sensor should not be too high to avoid the overload of the measuring system resulting from the maximum signal level.

10.2.6 The sensor shall provide a smooth response in the required frequency range. In the range from the lowest measured frequency to the highest measured frequency, the non-linear allowable deviation of the frequency-amplitude response of the sensor should be ±3 dB.

10.2.7 Sensors shall be firmly connected to the structure, avoiding loosening during vibration. The cable and signal line shall be fixed firmly. When installing a sensor with a special tool, the tool shall have sufficient stiffness to ensure its natural frequency is far greater than the maximum measured frequency after installation.

10.3 Detection points layout

10.3.1 The points for vibration response detection shall be arranged where the vibration response is relatively large. The points layout for dynamic characteristics detection shall be determined according to the structural type.

10.3.2 For vibration response and dynamic characteristics detection, sensors shall be arranged where the vibration amplitude is large, along the radial direction of steel penstocks section. 3 or 4 detection points are selected, and each detection point is provided with 1, 2 or 3 unidirectional vibration sensors which are mutually perpendicular along the vibration direction. When only 1 sensor is arranged, the vibration-detecting direction shall coincide with the direction of structural vibration.

10.3.3 When detecting vibration modes, sensors shall be arranged along the axis of penstocks, and the distribution of detection points shall entirely reflect the vibration modes.

10.4 Detection of vibration response

10.4.1 When detecting the vibration response of generating units, the vibration response corresponding to normal operating condition and load rejection condition shall be detected respectively.

10.4.2 During detection, the information related to the detection, such as upstream and downstream water levels, working conditions and running time, shall be recorded.

10.4.3 The vibration response data shall include the vibration amplitude, phase-frequency relation, and response amplitude of main frequencies.

10.5 Detection of dynamic characteristics

10.5.1 Vibration exciter, pulse exciter and ambient excitation may be used in the detection of dynamic characteristics. Vibration displacement, velocity, acceleration and frequency response are detected by sensors.

10.5.2 The detection results of dynamic characteristics shall include the first five natural frequencies, damping ratio and vibration modes of penstocks.

11 Check calculation

11.1 The check calculation method shall comply with NB/T 35056.

11.2 The result requirements for the check calculation are as follows:
  a) Under inspecting condition, the strength and stability of penstocks shall be checked;
  b) Under design or check condition, the strength and stability of penstocks shall be checked.

11.3 The load combination of the check calculation shall meet the design requirements. When the operating condition of steel penstocks changes, the load combination shall be redefined according to the actual situations.

11.4 When the materials of penstocks are consistent with those specified in the design document, check calculation shall be performed accordingly. When the materials are inconsistent with those specified in the design document, the check calculation shall be performed after the materials are tested and confirmed.

11.5 In the check calculation, the wall thickness of penstocks and the cross-sectional dimensions of main load-bearing components shall be taken from the measured values.

11.6 Baseing on the specifications in NB/T 35056, the adoption of evaluation indicators shall consider the influence of running time. The time influence coefficient shall be determined as follows:
  a) For the penstocks that run less than 10 years, the time influence coefficient shall be taken as 1.00;
  b) For the penstocks that run 10 to 30 years, the time influence coefficient shall be taken as 0.95;
  c) For the penstocks that run more than 30 years, the time influence coefficient shall be taken as 0.90.

12 Safety evaluation

12.1 The safety grades of steel penstocks can be classified into safe, basically safe, and unsafe.

12.2 The penstocks meeting the following conditions shall be assessed as safe:
  a) All items of patrol inspection meet the requirements;
  b) All items of appearance inspection meet the requirements;
  c) The degree of corrosion is not severer than Grade B;
  d) Materials meet the design requirements;
  e) The quality of Class I and Class II welds meet the requirements of relevant standards;

- f) The strength and stability against external pressure under design condition meet the requirements of relevant standards. The strength under load rejection of generating unit meets the requirements of relevant standards;
- g) No obvious vibration occurs during operation.

12.3 The penstocks meeting the following conditions shall be assessed as basically safe:
- a) All items of patrol inspection meet the requirements;
- b) All items of appearance inspection meet the requirements;
- c) The degree of corrosion is not severer than Grade C;
- d) Materials meet the design requirements;
- e) The quality of the Class I and Class II welds meets the requirements of relevant standards;
- f) The strength and stability against external pressure under design condition basically meet the requirements of relevant standards. The strength under load rejection of generating unit basically meets the requirements of relevant standards. The maximum value shall not exceed 105 % of the resistance limit;
- g) Obvious vibration occurs during operation, but it does not affect the operation safety of penstocks concluded from testing and analysis.

12.4 The penstocks that fail to meet any of the conditions as specified in Item 12.3 of this code shall be assessed as unsafe.

13 Inspection report

13.1 The inspection body shall provide the inspection report of penstocks to the entrusting party. The cover of the report shall be stamped with the CMA mark of the inspection body.

13.2 The inspection report shall include:
- a) Project overview and the operation, maintenance and repair of penstocks;
- b) Field inspection results of penstocks;
- c) Check calculation results of penstocks;
- d) Safety evaluation of penstocks;
- e) Conclusions and recommendations.

Annex A
(informative)
Patrol inspection record form

A.1 See Form A.1 for patrol inspection records of exposed penstocks.

Form A.1 Patrol inspection records of exposed penstocks

Name of power station			Penstock number	
Date		Temperature	°C	Weather	
Patrol inspection items	Patrol inspection records			
Deformation and displacement of pipe body	
Seepage of expansion joints	
Stroke of expansion joints	
Sealing performance of manholes and other openings	
Sealing performance of pipe joints	
Running status of vent or air valve	
Running status of drainage facilities	
Displacement of support piers	
Displacement of anchor blocks	
Settlement of support piers	
Settlement of anchor blocks	
Movability and lubrication of supports	
Sealing of the connection between inlet valve and penstocks	
Others	
Recorded by		Checked by		

A.2 See Form A.2 for patrol inspection records of embedded penstocks.

Form A.2   Patrol inspection records of embedded penstocks

Name of power station			Penstock number	
Date		Temperature	°C	Weather	
Patrol inspection items	Patrol inspection records			
Seepage along embedded penstocks				
Running status of groundwater drainage facilities				
Sealing performance of manholes and other openings				
Sealing performance of pipe joints				
Running status of relevant monitoring instruments				
Damage status of relevant monitoring instruments				
...				
Others				
Recorded by		Checked by		

A.3 See Form A.3 for patrol inspection records of embedded penstocks in dam.

Form A.3 Patrol inspection records of embedded penstocks in dam

Name of power station				Penstock number	
Date		Temperature	°C	Weather	
Patrol inspection items	Patrol inspection records				
Cracking of downstream dam surface					
Seepage of downstream dam surface					
Displacement of downstream dam surface					
Deformation of downstream dam surface					
...					
Others					
Recorded by			Checked by		

A.4 See Form A.4 for patrol inspection records of steel-lined reinforced concrete penstocks.

Form A.4   Patrol inspection records of steel-lined reinforced concrete penstocks

Name of power station				Penstock number	
Date		Temperature	°C	Weather	
Patrol inspection items					
Circumferential cracking in the upper bend section					
Circumferential cracking in the oblique straight section					
Axial cracking in the oblique straight section					
Seepage of expansion joints					
Stroke of expansion joints					
Sealing performance of manholes and other openings					
Sealing performance of pipe joints					
Running status of relevant monitoring instruments					
Damage status of relevant monitoring instruments					
...					
Others					
Recorded by			Checked by		

Annex B
(informative)
Appearance inspection record form

Form B.1 Appearance inspection records

Name of power station			Penstock number	
Inspection items			Inspection records	
Appearance of the pipe body	Deformation such as pipe wall indentation and bulging			
	Corrosion of pipe wall			
	Surface defects of main load-bearing welds			
	Damage and deformation of stiffener rings			
	Status of grouting holes and surrounding areas			
	Damage and deformation of manhole covers			
Appearance of supporting rings and supports	Deformation and damage			
	Running and contact status of supports			
	Corrosion status			
Appearance of the expansion joints	Deformation and damage			
	Running status			
	Corrosion status			
Appearance of connections	Connections between penstocks and concrete structures such as anchored blocks, dam and wall			
	Connections between penstocks and stiffener rings, supporting rings, or expansion joints			
	Connection between penstocks and inlet valve			
	Cavitation of concrete around penstocks			
Inspected by			Checked by	